Graphic Novel Series On Fractions

Book - 1

Finding Fractions!

Introduction To Fractions Using Everyday Examples

CuriousDots

Many thanks to the children, parents and teachers for adopting and experiencing the **SAPID learning experience**™ and providing valuable feedback.

Print Edition ISBN-13: 978-0-9893751-2-2

Dear Reader,

Welcome to a graphic and fun way of learning mathematics! The **SAPID learning experience**™ was created by a mathematician and a mother, when she was looking for a fun way to introduce pre-algebra topics, such as fractions and decimals, to her daughter. Though there are many approaches and books that teach mathematics, the goal seems to be to get a child ahead of the topics and levels allocated for their age. Children go through the grind of doing pages and pages of the same concept over and over again. While for anyone, let alone children, this gets boring and frustrating, it has the danger of a child developing an aversion to mathematics forever. We see many adults with a mindset that mathematics is not their cup of tea.

Our vision is to make learning fun! And when one has fun, there are no boundaries or levels or age. How far one can go is unlimited!   Our mission is to nurture and expand the creativity in a child. Not to set boundaries through structured approaches that diminish their fun and creativity.

The **SAPID learning experience**™ first introduces a concept through a story using a graphic novel format. This is followed by more stories to understand various aspects of a concept. The stories are short, mostly built around animal characters and are interactive, weaving practice and experiencing of concepts as the story unfolds. It enhances children's creativity and innovation by engaging them in art, lets them put on their detective hat and do some investigation and share their thinking on various solutions to a given situation.

When you are learning a concept such as fractions, many related topics are explored. A concept such as a simple fraction cannot happen in isolation in real life. Other topics such as graphs or measurements or geometry are intertwined naturally and together the stories are shaped into everyday scenarios, taking place in a kitchen or at a game or a playdate. At the heart of these books is the well-known philosophy that stories stick. Revisiting topics are important and stories are easy to recollect.

While some amount of practice is important, this approach will drastically reduce the need for repeated practice on paper. But practice will happen spontaneously as you weave mathematics into everyday life.

Without feedback from several children, parents and teachers this book would not have happened. Please write to us at curiousdots@gmail.com with feedback and suggestions to make learning fun.

CuriousDots team

CuriousDots

TABLE OF CONTENTS

How To Use This Book:

There are many ways to approach a book of this kind. One way would be to read the book end-to-end like a self-help manual. While this approach will benefit adults such as parents and teachers, for children this book needs to be used in a collaborative setting such as 1-on-1 interaction with a parent or educator or in a classroom setting.

Parents bond with their children by reading daily, especially during bedtime. This book can be used in a similar way. Open dialog and daily conversation about key concepts will make mathematics as much fun as other activities, such as reading and sports.

Children will benefit from reading each of the stories in a group setting or with a parent or an educator. They will naturally get involved in the stories and practice the exercises along with the story characters. Using the questions in the *STORY DISCUSSION* section, children can talk over each story and explore the topic from various angles.

The section *THEY ARE EVERYWHERE* provides tips to discover fractions in familiar everyday surroundings.

The section *CREATE YOUR OWN STORY* encourages the children's creativity, as they make up their own story with a set of keywords. If a child wants to get super creative and sketch a graphic novel, we have that covered as well! A few graphic novel templates are provided at the end of the book.

The entire book could be completed in a day or within four to five days.

Enjoy!

Snake's New Game!

ONCE UPON TIME THERE WAS A SNAKE NAMED MARISSA. SHE WAS EXCITED AS THE SUMMER VACATION WAS OVER AND IT WAS HER FIRST DAY AT SCHOOL.

School!!!

MARISSA WENT TO HER CLASSROOM AND WAS EXCITED TO MEET ALL HER FRIENDS. WHILE THEY WERE ALL CHATTING, A NEW SNAKE ENTERED THE CLASSROOM.

Hi Marissa!

Hi Rainbow!

How can I make her feel welcome?

THE CLASS BECAME QUIET FOR A FEW SECONDS AND THE NEW SNAKE WENT AND SAT IN A CORNER.

JUST THEN, THEIR TEACHER MR. SALVADOR ENTERED AND GREETED THE CLASS. HE INTRODUCED THE NEW STUDENT AS SPRINGY.

Everyone, this is Springy. She is joining our class this year!

I am sure all of you will make her feel very welcome.

MARISSA INVITED SPRINGY TO SIT NEXT TO HER. SPRINGY FELT GOOD AND WENT AND SAT NEXT TO MARISSA.

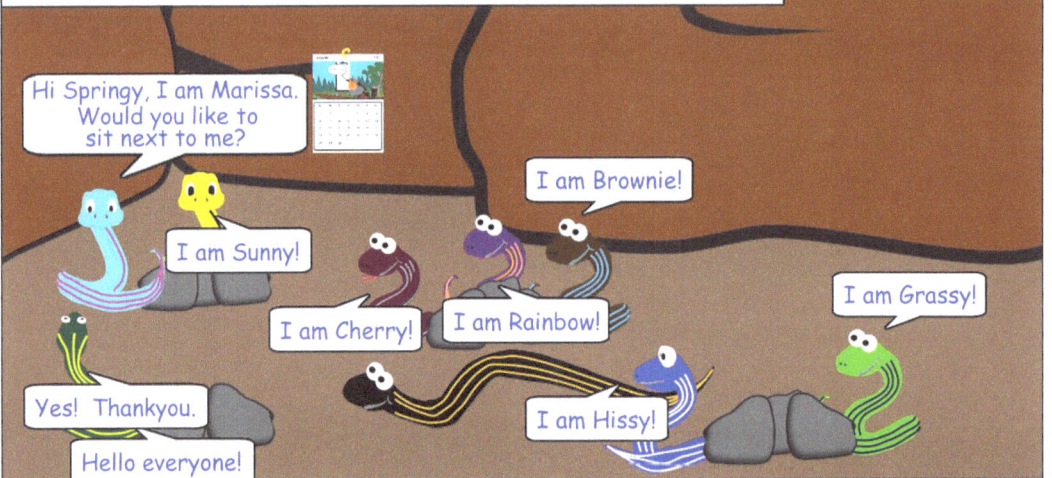

Hi Springy, I am Marissa. Would you like to sit next to me?

I am Brownie!

I am Sunny!

I am Grassy!

I am Cherry!

I am Rainbow!

Yes! Thankyou.

I am Hissy!

Hello everyone!

THE CLASS SPENT THEIR MORNING READING AND THEN WRITING IN THEIR JOURNAL. IT WAS NOW RECESS TIME! ALL THE SNAKES WENT OUTSIDE TO PLAY.

CuriousDots

Snake's New Game!

A SNAKE NAMED HISSY INTRODUCED THEM TO A NEW GAME. FIVE OF THE SNAKES INCLUDING MARISSA, SPRINGY AND HISSY GOT TOGETHER.

Can we all line up with our bodies in parallel and with our tails in line?

When I say, three, two, one, go, we all start moving and slithering at the same time.

We will make a beautiful pattern!

MARISSA THOUGHT THIS WAS FUN AND COMMENTED...

This is cool!

Wow! Springy! you are really tall!

Oh, I guess I am!

I think I am only half your length!

Maybe...

I am smaller than Marissa.

So I must be less than a half of you, Springy!

How much would that be?

I am not sure how to express something that is less than a half.

Maybe we can ask Mr. Salvador when we get back to our class.

Mr. Salvador! We think Marissa is half as tall as Springy, and I know I am smaller than Marissa.

Is there a way I can compare myself to Springy in length?

What would be less than a half?

Excellent question! Now that leads us to learn something new about numbers. Can one of you spell half?

FULL

HALF

H-A-L-F

Another way to express half!?

Perfect! There is another way to express a half. Does anyone know what that is?

THE CLASSROOM WAS QUIET.

MR. SALVADOR EXPLAINED THAT THERE WAS A NUMERIC WAY OF REPRESENTING A HALF. LIKE ONE CAN BE WRITTEN AS A WORD "O-N-E" AND CAN ALSO BE WRITTEN AS A NUMBER 1, HALF CAN ALSO BE WRITTEN AS A NUMBER.

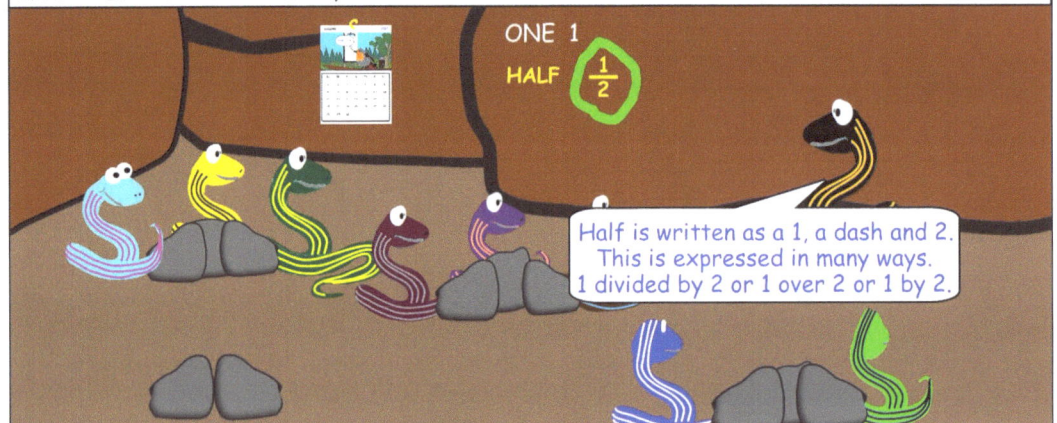

ONE 1

HALF $\frac{1}{2}$

Half is written as a 1, a dash and 2. This is expressed in many ways. 1 divided by 2 or 1 over 2 or 1 by 2.

CuriousDots

Snake's New Game!

MR. SALVADOR DREW TWO STRAIGHT LINES.

ONE 1
HALF ½

6 inches   Marissa
12 inches   Springy

Why is half a 1 by 2, Mr. Salvador?

Let's say this is Marissa's length and this one is Springy's length. Now, why do we say Marissa is half of Springy?

Springy is 12 inches which is twice as long as Marissa's length of 6 inches. In other words, Marissa is one part out of two equal parts!

MR. SALVADOR EXPLAINED THAT SINCE 1 BY 2 IS USED FREQUENTLY, INSTEAD OF SAYING, FOR EXAMPLE, 1 BY 2 OF AN APPLE, THE WORD HALF WAS INVENTED.

ONE 1
HALF ½
QUARTER ¼
THREE-QUARTER

6 inches   Marissa
Springy

Great explanation, Grassy!

There are a couple more words like this. Quarter is used to express 1 by 4 that is, 1 part of 4 parts. Three-quarter is used to express 3 by 4, that is 3 parts of 4 parts.

Cool! I did not know there were more numbers other than 0, 1, 2, 3 and so on.

You are on the right track, Hissy!

ONE 1
HALF ½
QUARTER ¼
THREE-QUARTER ¾

6 inches   Marissa
12 inches   Springy

Aside from the whole numbers, there is always a need to express something that is a fraction or a part of a thing. Like half an apple!

And such numbers which are not whole numbers have a special name! These numbers are called fractions!

ONE 1
HALF 1/2
6 inches Marissa
12 inches Springy
That makes sense! But, if I am half of Springy, then would I say Springy is two parts out of two parts that is, two by two? Now that sounds strange!
Wow! All of you are asking great questions!
You are correct that Springy's length can be represented as two by two or better as one which is the whole length.
One important thing. For fractions to exist, all the parts needs to be equal.
ONE 1
HALF 1/2
QUARTER 1/4
THREE-QUARTER 3/4
6 inches Marissa
12 inches Springy
Now, can you all form small groups and and discuss how to express Hissy's length as a fraction of Springy?
But what is Hissy's length?
THREE-QUARTER 3/4
I am glad you asked! Let us assume that she is three inches.

CAN YOU HELP HISSY AS SHE COMPARES HER LENGTH TO SPRINGY'S? SEE NEXT PAGE FOR THE CLASS WORKSHEET.

12 inches

1

Springy is the tallest! She is twelve inches in length!

6 inches

$\frac{1}{2}$

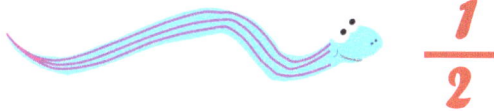

Six is half of twelve. So Marissa is half of Springy!

3 inches

Three is _____ of twelve.

So Hissy is _____ of Springy!

Hissy's length in fraction compared to Springy's length

12 inches

1

Springy is the tallest! She is twelve inches in length!

6 inches

$\dfrac{1}{2}$

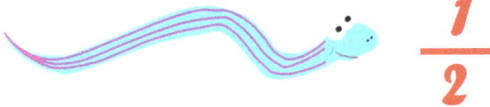

Six is half of twelve. So Marissa is half of Springy!

3 inches

$\dfrac{1}{4}$

Three is _____quarter_____ of twelve.

So Hissy is _____quarter_____ of Springy!

CuriousDots

KITCHEN

We make the best...

Can we take a slice of pizza for Coco? Can you please prepare another slice?

...eeks special

That is very thoughtful of you, Hissy. I do not need to make another slice. Since I made a small cheese pizza which has four slices and you only had one slice, there are three remaining slices . I will pack one for Coco.

Wet Paint! Do not touch, please!

ELLIE STARTED WALKING TOWARDS THE KITCHEN AND SUDDENLY STOPPED.

KITCHEN

We make the best pizza!

Wait a minute! Did you know that a pizza or a pie are great examples for fractions?

But a pizza is circular in shape!

...cial

How can a pizza be represented as a fraction?

Wet Paint! Do not touch, please!

HISSY WAS SURPRISED THAT ELLIE KNEW ABOUT FRACTIONS AND WONDERED HOW PIZZAS COULD BE REPRESENTED AS A FRACTION.

This weeks special

Why don't you come with me and I will show you how pizzas are a great example of fractions?

Wet Paint! Do not touch, please!

HISSY TURNED TO HER GRANDMOTHER AND ASKED HER PERMISSION. HER GRANDMOTHER AGREED.

CuriousDots

ELLIE AND HISSY WALKED INTO THE KITCHEN. HISSY THOUGHT THE KITCHEN SMELLED HEAVENLY. MAC, ANOTHER ELEPHANT WAS BUSY CUTTING A LARGE PIZZA.

Do you know that the first rule for fractions is that all parts need to be equal?

Yes, my teacher Mr. Salvador mentioned it in our class.

HISSY WAS AGAIN SURPRISED BY ELLIE'S KNOWLEDGE.

MAC STACKED UP THE SLICES OF A PIZZA ONE ON TOP OF THE OTHER. HISSY REALIZED THAT THEY WERE ALL THE SAME SIZE. SHE QUICKLY UNDERSTOOD WHY A PIZZA WAS A GREAT EXAMPLE OF FRACTION.

So a slice of pizza can be represented as one by eight?

1/8?

Exactly! Since in a large pizza there are 8 slices, one slice can be represented as one eighth of a pizza, two slices would be two eighth of a pizza and so on. Eight slices would be eight eighth of a pizza or the whole pizza.

But why can't I just say one slice or two slices?

There is nothing wrong in saying one slice or two slices. But when you say one eighth of a pizza, you have additional information on the pizza. That is, it is not just one slice but one slice out of the total of eight slices.

Let us look at some examples using pizza slices.

Page Intentionally Left Blank

ELLIE PULLED OUT SOME MODELS OF PIZZA SLICES. BELOW ARE THE DIFFERENT WAYS SHE DISPLAYED THE PIZZA SLICES AND ASKED HISSY TO ASSOCIATE A NUMBER TO EACH OF THE DISPLAY. WOULD YOU LIKE TO TRY? HISSY'S RESPONSES ARE SHOWN IN THE NEXT PAGE.

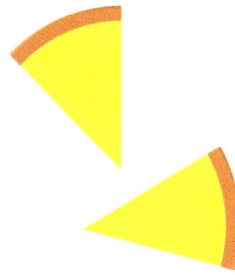

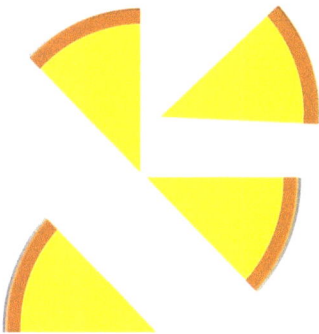

HERE ARE THE NUMBERS HISSY ASSOCIATED TO EACH OF THE DISPLAY.

$$\frac{2}{8}$$

2

4

$$\frac{3}{8}$$

CuriousDots

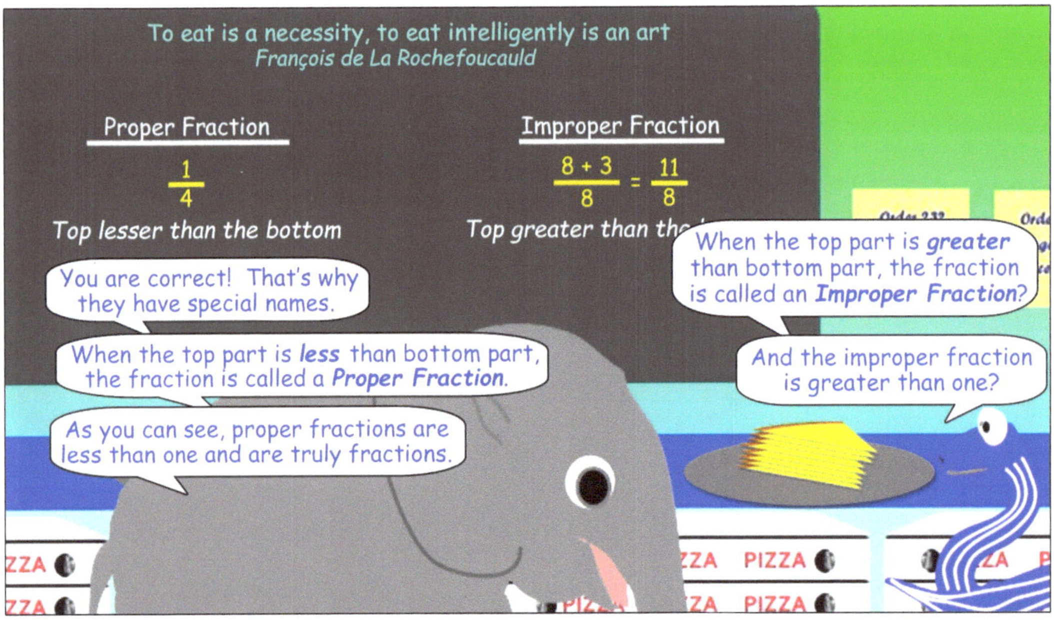
To eat is a necessity, to eat intelligently is an art
François de La Rochefoucauld
Proper Fraction
1/4
Top lesser than the bottom
Improper Fraction
(8 + 3)/8 = 11/8
You are correct! That's why they have special names.
When the top part is less than bottom part, the fraction is called a Proper Fraction.
As you can see, proper fractions are less than one and are truly fractions.
When the top part is greater than bottom part, the fraction is called an Improper Fraction?
And the improper fraction is greater than one?
PIZZA

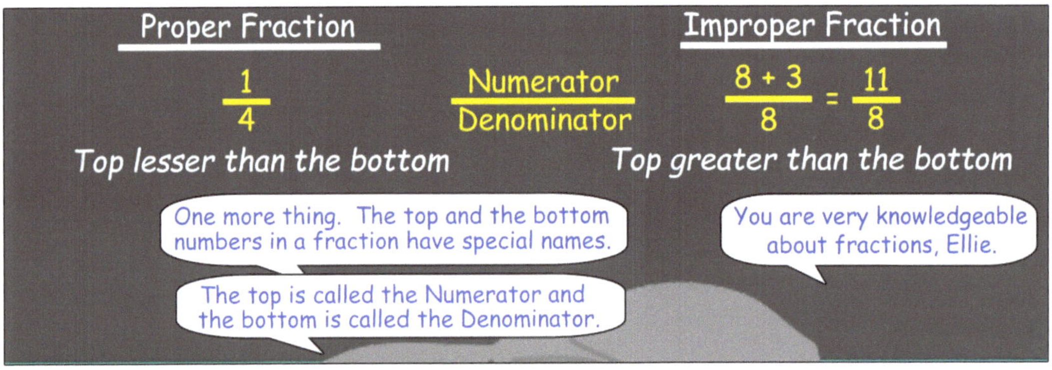
Proper Fraction
1/4
Numerator/Denominator
Improper Fraction
(8 + 3)/8 = 11/8
Top lesser than the bottom
Top greater than the bottom
One more thing. The top and the bottom numbers in a fraction have special names.
The top is called the Numerator and the bottom is called the Denominator.
You are very knowledgeable about fractions, Ellie.

To eat is a necessity, to eat intelligently is an art
Proper Fraction
Improper Fraction
Numerator/Denominator
Top lesser than the bottom
Top greater than the bottom
If a recipe calls for one third of salt and if one does not know what that means, the food will end up either too salty or with less salt! That will not taste good.
You will find fractions everywhere in this kitchen.
Math is applied everywhere, especially in cooking. Cooking is one of the best scientific experiments and one gets to eat it too!
Why don't you go around the kitchen and see what you can find.
PIZZA

Cheese Pizza Ingredients:
Pizza crust
1/2 - 3/4 cup pizza sauce
1 or 1/2 cups shredded mozzarella cheese
2-3 Tablespoons grated Parmesan cheese
Ground basil, to taste

Tomato Sauce Ingredients:
1 (6 ounce) can tomato paste
6 fluid ounces warm water (110 F)
3 tablespoons grated Parmesan cheese
1 teaspoon minced garlic
2 tablespoon honey
3/4 teaspoon onion powder
1/4 teaspoon dried oregano
1/4 teaspoon dried basil
1/4 teaspoon ground black pepper
Salt to taste

HISSY LOOKED AT THE RECIPES ON THE WALL AND SAID THAT THEY WERE FULL OF FRACTIONS. THEN SHE SAW A MEASURING CUP WITH FRACTIONS! SEVERAL MEASURING SPOONS WITH FRACTIONS! SHE ALSO SAW A YUMMY SQUARE CAKE CUT INTO NINE PIECES AND THEY WERE EQUAL.

Can I have one ninth of the cake please!

Of Course! Now let's go.

Thanks Ellie. Fractions are everywhere! I am going to share this with my class.

HER GRANDMOTHER HAD ALMOST FINISHED THE PIZZA. HISSY SHARED HER EXCITEMENT ABOUT THE THINGS SHE HAD LEARNED IN THE KITCHEN.

We make the best pizza!

So what example of fractions do you see around here?

This weeks special

KITCHEN

Wet Paint! Do not touch, please!

HISSY LOOKED AROUND AND THERE WERE ONLY TABLES, CHAIRS AND SOME SALT AND PEPPER SHAKERS ON EACH TABLE. SHE OBSERVED THAT TWO TABLES OF THE SEVEN TABLES HAD THE SIGN "WET PAINT. DO NOT TOUCH PLEASE!"

CuriousDots

THEY THANKED ELLIE AND FED THE PIZZA TO COCO. HISSY LOOKED AROUND AND SAW THERE WERE FIVE CARS BUT ONLY THEIR CAR WAS BLUE.

Grandma! One fifth of the cars is blue and that happens to be ours!

That is correct! Now, how much pizza is left here?

THE BOWL WAS EMPTY AND COCO WAS WAGGING HER TAIL HAPPILY.

Zero by four?

Correct! And zero by four is actually zero!

Let us drive Grandma! I want to find a lot of examples of fractions when we are driving!

I want to write and share them with my friends.

Page Intentionally Left Blank

CuriousDots

HERE ARE THE DETAILS HISSY SHARED ABOUT THE FRACTIONS SHE CAME ACROSS AT THE PIZZA SHOP AND ON THEIR DRIVE BACK HOME. TRY OUT THE QUESTIONS! SEE NEXT PAGE FOR HER FRIENDS RESPONSES.

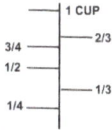

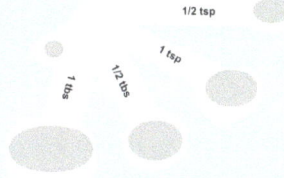

1 CUP
3/4
1/2
1/4
2/3
1/3

1/2 tsp
1 tsp
1 tbs
1/2 tbs

Cheese Pizza Ingredients:
Pizza crust
1/2 - 3/4 cup pizza sauce
1 or 1/2 cups shredded mozzarella cheese
2-3 Tablespoons grated Parmesan cheese
Ground basil, to taste

21 1/8 Rd

Grand Ave	1/4
Woodland Ave	1 1/2
END FREEWAY	1 1/4

Regular	$389\frac{9}{10}$
Plus	$409\frac{9}{10}$
Premium	$419\frac{9}{10}$
Diesel	$459\frac{9}{10}$

How many unique proper fractions and improper fractions did I find?

What are the names given to 1 and 4 in the fraction 1/4?

Fractions found in Ellie's pizza shop and while driving home

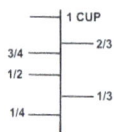

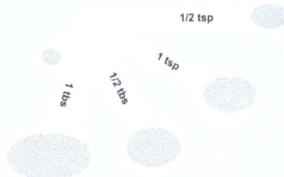

```
        1 CUP
            2/3
3/4
1/2
            1/3
1/4
```

1/2 tsp

1 tsp

1 tbs   1/2 tbs

Cheese Pizza Ingredients:
Pizza crust
1/2 - 3/4 cup pizza sauce
1 or 1/2 cups shredded mozzarella cheese
2-3 Tablespoons grated Parmesan cheese
Ground basil, to taste

21 1/8 Rd

Grand Ave	**1/4**
Woodland Ave	**1 1/2**
END FREEWAY	**1 1/4**

Regular	**389 9/10**
Plus	**409 9/10**
Premium	**419 9/10**
Diesel	**459 9/10**

How many unique proper fractions and improper fractions did I find?

Proper Fractions - 7   Improper Fractions - 0

What are the names given to 1 and 4 in the fraction 1/4?
1 is called the numerator. 4 is called the denominator.

They are everywhere...

Exploring Everyday Examples

Look for items that can be grouped as a whole, and the various categories that can be expressed as fractions. Remember, as Hissy's grandmother said, they need not be of equal size or shape. Below are few examples, to get you started.

1. Windows and doors in your house
2. Various colored blocks in a play set
3. T-shirts and colors
4. A particular donut in a bakery shop

Investigation

Here are some curious questions on fractions. Try answering them, and can you come up with more questions?

1. Do we really need fractions? Are whole numbers such as 0, 1, 2, 3... adequate?
2. Can the denominator be zero?
3. Can the numerator be zero?
4. Can both the numerator and denominator be the same number?
5. Why are the names proper and improper fractions? Are proper fractions really proper?

Can you find an improper fraction?

Finding an improper fraction out there is not easy. Here are a few places we explored.

1. While driving on a highway
2. In the kitchen
3. In a classroom

Discuss stories using the following questions as guidelines.

1. What is the predominant theme in the story? Around who is the story centered?
2. Where does the primary action take place?
3. How does the story get started? What is the initial incident?
4. Briefly describe the rising action of the story.
5. What is the high point, or climax, of the story?
6. Discuss the falling action or end of the story.
7. Was there a villain in the story? a hero? a dynamic character?
8. Does the story contain a single effect or impression for the reader? If so, what is it?

Look around you and create a story that includes fractions.

Use the space below to draft your own story with your own characters. The characters could even be you and your friends! Use some or all of the keywords listed below in your story.

If you want to draw and present your story as a graphic novel, use the templates at the end of the book.

You can even try using your script to enact it before a group or your class.

Keywords: Fraction, Numerator, Denominator, Proper fraction, Improper fraction

CuriousDots

Reproduce as many copies as needed

Reproduce as many copies as needed

CuriousDots

www.ingramcontent.com/pod-product-compliance
Lightning Source LLC
Chambersburg PA
CBHW052044190326
41520CB00002BA/183